Oxford University Press, Walton Street, Oxford OX2 6DP
Oxford New York
Athens Auckland Bangkok Bombay
Calcutta Cape Town Dar es Salaam Delhi
Florence Hong Kong Istanbul Karachi
Kuala Lumpur Madras Madrid Melbourne
Mexico City Nairobi Paris Singapore
Taipei Tokyo Toronto

and associated companies in
Berlin Ibadan

Oxford is a trade mark of Oxford University Press

ISBN 0 19 279537 6 (Hardback)
ISBN 0 19 272208 5 (Paperback)

First edition 1954
Reprinted 1956, 1957, 1960, 1962, 1964, 1965, 1967, 1968,
1969, 1972, 1975, 1982, 1983, 1984, 1985, 1987, 1989, 1990, 1991
Reprinted with new cover 1996
First published in paperback 1989
Reprinted 1990, 1991, 1992
Reprinted with new cover 1996

Printed in Hong Kong

Lavender's blue, dilly, dilly,
  Lavender's green;
When I am king, dilly, dilly,
  You shall be queen.

CONTENTS

Rock-a-bye, Baby

Rock-a-bye, baby
Hush-a-bye, baby
Grey goose and gander
Dance to your daddy
Ride a cock-horse
This is the way the ladies ride
Bye, baby bunting
Ring-a-ring o' roses
Two little dicky birds
Foxy's hole
Dance, Thumbkin, dance
Here sits the Lord Mayor
Walk down the path
Here's a ball for baby
Pat-a-cake, pat-a-cake, baker's man
Can you keep a secret?
Here are my lady's knives and forks
Warm hands, warm
Round and round the cornfield
This little pig went to market
Shoe the little horse
Sleep, baby, sleep

Rock-a-bye, baby,
   Thy cradle is green,
Father's a nobleman,
   Mother's a queen;
And Betty's a lady,
   And wears a gold ring;
And Johnny's a drummer,
   And drums for the king.

Hush-a-bye, baby,
    on the tree top,
When the wind blows,
    the cradle will rock;

When the bough breaks,
    the cradle will fall,
Down will come baby,
    bough, cradle and all.

Grey goose and gander,
Waft your wings together,
And carry the good king's daughter
Over the one strand river.

Dance to your daddy,
My little babby,
Dance to your daddy,
my little lamb;
You shall have a fishy
In a little dishy,
You shall have a fishy
when the boat
comes in.

Ride a cock-horse to Banbury Cross,
To see a fine lady upon a white horse;
With rings on her fingers and bells on her toes,
She shall have music wherever she goes.

This is the way the ladies ride,
  Tri, tre, tre, tree,
  Tri, tre, tre, tree;
This is the way the ladies ride,
  Tri, tre, tre, tre, tri-tre-tre-tree!

This is the way the gentlemen ride,
  Gallop-a-trot,
  Gallop-a-trot;
This is the way the gentlemen ride,
  Gallop-a-gallop-a-trot!

This is the way the farmers ride,
  Hobbledy-hoy,
  Hobbledy-hoy;
This is the way the farmers ride,
  Hobbledy hobbledy-hoy!

Bye, baby bunting,
   Daddy's gone a-hunting,
Gone to get a rabbit skin
   To wrap a baby bunting in.

Ring-a-ring o' roses,
A pocket full of posies,
A-tishoo, a-tishoo!
We all fall down.

Two little dicky birds
Sat upon a wall,
One named Peter,
The other named Paul.

Put your finger in Foxy's hole,
Foxy's not at home;
He's out at the back door
Picking at a bone.

Fly away Peter !
Fly away Paul !
Come back Peter,
Come back Paul.

Dance, Thumbkin, dance,
Dance, ye merrymen, every one ;
But Thumbkin, he can dance alone.

Dance, Foreman, dance...

Dance, Longman...

Dance, Ringman...

Dance, Littleman, dance,
Dance, ye merrymen, every one ;
But Littleman he can't dance alone.

Here sits the Lord Mayor,
Here sit his men,
Here sits the cock,
Here sits the hen,
Here sit the little chickens,
Here they run in,
Chinchopper, chinchopper,
                    chinchopper, chin.

Walk down the path,
  Knock at the door,
Lift the latch,
  Wipe your feet and walk in.

Here's a ball for baby,
    Big and soft and round;
Here is baby's hammer,
    See how he can pound!
Here are baby's soldiers,
    Standing in a row;
This is baby's music,
    Clapping, clapping so!
Here's a big umbrella
    To keep the baby dry;
    Here is baby's cradle,
    Rock a baby bye.

Pat-a-cake, pat-a-cake, baker's man,
　　Make me a cake as fast as you can:
Pat it and prick it, and mark it with B,
　　And toss it in the oven for baby and me.

Can you keep a secret?
　　I don't believe you can!
You mustn't laugh,
　　You mustn't cry,
But do the best you can.

Here are my lady's knives and forks,
Here is my lady's table,
Here is my lady's looking-glass,
And here is the baby's cradle.

Warm hands, warm,
The men have gone to plough;
If you want to warm your hands,
Warm your hands now.

Round and round the cornfield
Looking for a hare,
Where can we find one?
Right up there!

This little pig went to market,

This little pig stayed at home,

This little pig had roast beef,
And this little pig had none,

And this little pig cried, Wee, wee, wee!
All the way home.

Shoe the little horse,
   Shoe the little mare,
But let the little colt
   go bare, bare, bare.

Sleep, baby, sleep,
 Thy father guards the sheep;
Thy mother shakes the dreamland tree
 And from it fall sweet dreams for thee,
Sleep, baby, sleep.

Girls and Boys come out to Play

Girls and boys come out to play
Georgie Porgie
Little Boy Blue
Little Polly Flinders
Little Tommy Tucker
Lavender's blue
Simple Simon
Polly put the kettle on
Little Miss Muffet
I had a little nut-tree
Tom, Tom, the piper's son
Tommy Snooks and Bessie Brooks
There was an old woman who lived in a shoe
Jack and Jill
Jack be nimble
How many miles to Babylon?
Wee Willie Winkie
Little Bo-Beep
Little Jack Horner
Mary, Mary, quite contrary
Here am I, little Jumping Joan
A dillar, a dollar
Three children sliding on the ice
Curly Locks
Lucy Locket
Upon Paul's steeple

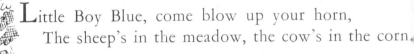

Little Boy Blue, come blow up your horn,
The sheep's in the meadow, the cow's in the corn.

Little Polly Flinders
    Sat among the cinders
Warming her pretty little toes.
    Her mother came and caught her,
And whipped her little daughter
    For spoiling her nice new clothes.

Where is the boy that looks after the sheep?
He's under the haycock fast asleep!

Little Tommy Tucker
    Sings for his supper;
What shall we give him?
    White bread and butter.
How shall he cut it,
    Without e'er a knife?
How can he marry
    Without e'er a wife?

Lavender's blue, dilly, dilly,
    Lavender's green;
When I am king, dilly, dilly,
    You shall be queen.
Call up your men, dilly, dilly,
    Set them to work,

Some to the plough, dilly, dilly,
  Some to the cart.
Some to make hay, dilly, dilly,
  Some to cut corn,
While you and I, dilly, dilly,
  Keep ourselves warm.

Simple Simon met a pieman,
　Going to the fair;
Says Simple Simon to the pieman,
　Let me taste your ware.
Says the pieman to Simple Simon,
　Show me first your penny.
Says Simple Simon to the pieman,
　Indeed I have not any.
Simple Simon went a-fishing
　For to catch a whale;
All the water he had got
　Was in his mother's pail.
Simple Simon went to look
　If plums grew on a thistle;
He pricked his finger very much,
　Which made poor Simon whistle.

Polly put the kettle on,
    Polly put the kettle on,
Polly put the kettle on,
    We'll all have tea.

Sukey take it off again,
    Sukey take it off again,
Sukey take it off again,
    They're all gone away.

Little Miss Muffet
　Sat on a tuffet,
　　Eating of curds and whey;

There came a great spider,
And sat down beside her,
　And frightened Miss Muffet away.

I had a little nut-tree, nothing would it bear
   But a silver nut-meg and a golden pear;
The King of Spain's daughter came to visit me,
   And all for the sake of my little nut-tree.

Tom, Tom, the piper's son,
  Stole a pig, and away did run.
The pig was eat, and Tom was beat,
  And Tom went roaring down the street.

As Tommy Snooks and Bessie Brooks
Were walking out one Sunday,
Says Tommy Snooks to Bessie Brooks,
Tomorrow will be Monday.

There was an old woman who lived in a shoe,
   She had so many children she didn't know what to do;
She gave them some broth without any bread,
   And whipped them all soundly and sent them to bed.

Jack and Jill went up the hill
    To fetch a pail of water;
Jack fell down
            and broke his crown,
    And Jill came tumbling after.

Then up Jack got
            and home did trot
    As fast as he could caper;
And went to bed
                to mend his head
    With vinegar
            and brown paper.

Jack be nimble,
　Jack be quick,
Jack jump over
　The candlestick.

How many miles to Babylon?
　Threescore miles and ten.
Can I get there by candle-light?
　Yes, and back again.
If your heels are nimble and light,
　You may get there by candle-light.

Wee Willie Winkie
　　　　　runs through the town,
Upstairs and downstairs
　　　　　in his nightgown,
Rapping at the window,
　　　　　crying through the lock,
Are the children in their beds?
For now it's eight o'clock.

Little Bo-Peep has lost her sheep,
    And can't tell where to find them;
Leave them alone, and they'll come home
    And bring their tails behind them.

Little Bo-Peep fell fast asleep,
    And dreamt she heard them bleating;
But when she awoke, she found it a joke,
    For they were still a-fleeting.

Then up she took her little crook,
    Determined for to find them,
She found them indeed, but it made her heart bleed
    For they'd left all their tails behind 'em.

It happened one day as Bo-Peep did stray
    Into a meadow hard by,
There she espied their tails side by side,
    All hung on a tree to dry.

She heaved a sigh and wiped her eye,
    Then went o'er hill and dale,
And tried what she could, as a shepherdess should,
    To tack to each sheep its tail.

Little Jack Horner
Sat in a corner,
Eating a Christmas pie;
He put in his thumb
And pulled out a plum,
And said, What a good
boy am I!

Mary, Mary, quite contrary,
How does your garden grow?
With silver bells and cockle-shells,
And pretty maids all in a row.

Here am I, little Jumping Joan.
When nobody's with me,
I'm always alone.

A dillar, a dollar,
A ten o'clock scholar.
What makes you come so soon?
You used to come at ten o'clock,
But now you come at noon.

Three children sliding on the ice
    All on a summer's day
As it fell out, they all fell in,
    The rest they ran away.

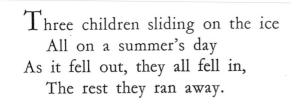

Now, had these children been at home,
    Or sliding on dry ground,
Ten thousand pounds to one penny,
    They had not all been drowned.

You parents all that children have,
    And you that have got none,
If you would have them safe abroad,
    Pray keep them safe at home.

Curly Locks, Curly Locks,
          wilt thou be mine?
Thou shalt not wash dishes,
          nor yet feed the swine,
But sit on a cushion
          and sew a fine seam,
And feed upon strawberries,
          sugar and cream.

Lucy Locket lost her pocket,
Kitty Fisher found it.
There was not a penny in it,
Only ribbon round it.

Upon Paul's steeple stands a tree
As full of apples as may be,
The little boys of London town
They run with hooks to pull them down;
And then they run from hedge to hedge
Until they come to London Bridge.

Old Mother Goose

Old Mother Goose
When good King Arthur ruled this land
The grand old Duke of York
Rub a dub dub
Tweedledum and Tweedledee
Jack Sprat
There was a man of Thessaly
One misty, moisty morning
Cross-patch
The Queen of Hearts
There was an old woman toss'd up in a basket
The man in the moon
Humpty Dumpty
Peter, Peter, pumpkin-eater
When I was a bachelor
If I'd as much money as I could spend
I saw three ships come sailing by
London Bridge is broken down
Dr Faustus
Peter Piper
The man in the wilderness
There was an old woman, and what do you think?
Old King Cole
Doctor Foster
There was an old woman lived under a hill
There was a crooked man
Diddle, diddle, dumpling
Come, let's to bed, says Sleepy-head
The jolly miller
Hark, hark, the dogs do bark
Solomon Grundy
Three wise men of Gotham
Jack-a-Nory
There was a little man

Old Mother Goose, when
    She wanted to wander,
Would ride through the air
    On a very fine gander.

Mother Goose had a house,
    'Twas built in a wood,
Where an owl at the door
    For sentinel stood.

She had a son Jack,
    A plain-looking lad,
He was not very good,
    Nor yet very bad.

She sent him to market,
  A live goose he bought,
Here, mother, says he,
  It will not go for nought.

Jack's goose and her gander
  Grew very fond;
They'd both eat together,
  Or swim in one pond.

Jack found one morning,
  As I have been told,
His goose had laid him
  An egg of pure gold.

When good King Arthur ruled
this land,
He was a goodly king;
He stole three pecks of
barley meal,
To make a bag-pudding.

A bag-pudding the king did
make,
And stuff'd it well with
plums.
And in it put great lumps of
fat,
As big as my two thumbs.

The king and queen did eat
thereof,
And noblemen besides;
And what they could not eat
that night,
The queen next morning
fried.

O, the grand old Duke of York,
    He had ten thousand men;
He marched them up to the top of the hill
    And he marched them down again!
When they were up, they were up,
    And when they were down, they were down,
And when they were only half way up,
    They were neither up nor down.

Rub a dub dub,
    Three men in a tub,
And who do you think they be?
    The butcher, the baker,
The candlestick-maker;
    Turn 'em out knaves all three!

Tweedledum and Tweedledee
    Agreed to have a battle,
For Tweedledum said Tweedledee
    Had spoiled his nice new rattle.
Just then flew by a monstrous crow,
    As big as a tar-barrel,
Which frightened both the heroes so
    They quite forgot their quarrel.

Jack Sprat could eat no fat,
His wife could eat no lean:
And so betwixt them both, you see,
They lick'd the platter clean.

There was a man of Thessaly,
And he was wond'rous wise;
He jump'd into a quickset hedge,
And scratch'd out both his eyes.
But when he saw his eyes were out,
With all his might and main
He jump'd into another hedge
And scratched 'em in again.

One misty, moisty morning
    When cloudy was the weather,
There I met an old man
    Clothed all in leather;
Clothed all in leather,
    With cap under his chin.
How do you do, and how do you do,
    And how do you do again?

Cross-patch,
    Draw the latch,
Sit by the fire and spin;
    Take a cup,
And drink it up,
    Then call your neighbours in.

The Queen of Hearts,
　　She made some tarts,
All on a summer's day;
　　The Knave of Hearts,
He stole those tarts,
　　And took them clean away.

The King of Hearts
　　Called for the tarts,
And beat the Knave full sore,
　　The Knave of Hearts
Brought back the tarts,
　　And vowed he'd steal no more.

There was an old woman toss'd up in a basket
    Nineteen times as high as the moon;
Where she was going I couldn't but ask it,
    For in her hand she carried a broom.

Old woman, old woman, old woman, quoth I,
    O wither, O wither, O wither, so high?
To brush the cobwebs off the sky!
    Shall I go with thee? Ay, by-and-by.

The man in the moon,
　　Came tumbling down,
And asked his way to Norwich;
　　He went by the south,
And burnt his mouth
　　With supping cold pease-porridge.

Humpty Dumpty sat on a wall;
Humpty Dumpty had a great
fall;
All the king's horses and all the
king's men
Couldn't put Humpty Dumpty
together again.

Peter, Peter, pumpkin-eater,
   Had a wife and couldn't keep her,
He put her in a pumpkin shell
   And there he kept her very well.

When I was a bachelor, I lived by myself,
And all the bread and cheese I got,
I put upon the shelf;
The rats and the mice did lead me such
a life,
That I went to London
to get myself a wife;

The streets they were so broad, and the
lanes they were so narrow,
I couldn't get my wife home
without a wheelbarrow;
The wheelbarrow broke,
my wife got a fall,
Down fell wheelbarrow,
little wife,
and all!

If I'd as much money as I could spend,
   I never would cry old chairs to mend;
Old chairs to mend, old chairs to mend,
   I never would cry, old chairs to mend.

If I'd as much money as I could tell,
   I never would cry young lambs to sell;
Young lambs to sell, young lambs to sell,
   I never would cry young lambs to sell.

I saw three ships come sailing by,
  Come sailing by, come sailing by;
I saw three ships come sailing by,
  On New Year's Day in the morning.

And what do you think was in them then,
  Was in them then, was in them then?
And what do you think was in them then,
  On New Year's Day in the morning.

Three pretty girls were in them then,
  Were in them then, were in them then;
Three pretty girls were in them then,
  On New Year's Day in the morning.

And one could whistle, and one could sing,
  And one could play on the violin —
Such joy there was at my wedding,
  On New Year's Day in the morning.

London Bridge is broken down,
   Dance o'er my Lady Lee,
London Bridge is broken down
   With a gay lady.
How shall we build it up again?
   Dance o'er my Lady Lee, etc.
Build it up with silver and gold,
   Dance o'er my Lady Lee, etc.
Silver and gold will be stole away,
   Dance o'er my Lady Lee, etc.
Build it up with iron and steel,
   Dance o'er my Lady Lee, etc.
Iron and steel will bend and bow,
   Dance o'er my Lady Lee, etc.
Build it up with wood and clay,
   Dance o'er my Lady Lee, etc.
Wood and clay will wash away,
   Dance o'er my Lady Lee, etc.
Build it up with stone so strong,
   Dance o'er my Lady Lee,
Huzza! 'twill last for ages long,
   With a gay lady.

Doctor Faustus was a good man,
     He whipped his scholars now and then;
When he whipped them he made them dance
     Out of England into France,
Out of France into Spain,
     And then he whipped them back again!

Peter Piper picked a peck of pickled pepper,
     A peck of pickled pepper Peter Piper picked.
If Peter Piper picked a peck of pickled pepper,
     Where's the peck of pickled pepper Peter Piper
                    picked?

The man in the wilderness asked me
  How many strawberries grew in the sea.
I answered him as I thought good,
  As many as red herrings grew in the wood.

There was an old woman, and what do you think?
  She lived upon nothing but victuals and drink:
Victuals and drink were the chief of her diet,
  Yet this little old woman could never keep quiet.

Old King Cole
  Was a merry old soul,
And a merry old soul was he;
  He called for his pipe,
And he called for his bowl,
  And he called for his fiddlers
          three.

Now every fiddler, he had a fiddle,
  And a very fine fiddle had he;
Twee tweedle dee, tweedle dee, went the
                  fiddlers.
    Oh, there's none so rare,
As can compare
    With King Cole and his fiddlers three!

Doctor Foster went to Glo'ster
  In a shower of rain;
He stepped in a puddle, right up to his middle,
  And never went there again.

There was an old woman
  Lived under a hill,
And if she's not gone
  She lives there still.

There was a crooked man, and he went a crooked mile,
He found a crooked sixpence against a crooked stile:
He bought a crooked cat, which caught a crooked mouse,
And they all lived together in a little crooked house.

Diddle, diddle, dumpling, my son John
Went to bed with his trousers on;
One shoe off, the other shoe on,
Diddle, diddle, dumpling, my son John.

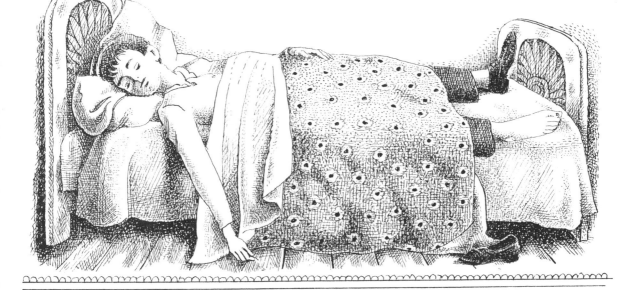

Come, let's to bed,
  Says Sleepy-head;
Tarry a while, says Slow;
  Put on the pan,
  Says Greedy Nan,
    Let's sup before we go.

There was a jolly miller once
    Lived by the river Dee,
He worked and sang from morn till night,
    No lark more blithe than he.
And this the burden of his song
    For ever used to be,
I care for nobody, no! not I,
    And nobody cares for me.

Hark, hark, the dogs do bark,
  The beggars are coming to town.
Some in rags, and some in tags,
  And some in velvet gowns.

Solomon Grundy,
  Born on a Monday,
Christened on Tuesday,
  Married on Wednesday,
Took ill on Thursday,
  Worse on Friday,
Died on Saturday,
  Buried on Sunday:
That is the end
  Of Solomon Grundy.

HERE
LIES
Solomon
Grundy

Three wise men of Gotham
   Went to sea in a bowl;
And if the bowl had been stronger,
   My tale had been longer.

I'll tell you a story
   About Jack-a-Nory,
And now my story's begun:
   I'll tell you another
About Jack and his brother,
   And now my story's done.

There was a little man,
    And he had a little gun,
And his bullets were made of lead, lead, lead;
He went to the brook
    And he saw a little duck,
And he shot it right through the head, head, head.

He carried it home
　　To his old wife Joan,
And bid her a fire for to make, make, make,
To roast the little duck
　　He had shot in the brook,
And he'd go and fetch her the drake, drake, drake.

The Lion and the Unicorn

The lion and the unicorn
Ladybird, ladybird
I had a little pony
Leg over leg
Little Robin Redbreast
Goosey, goosey, gander
I saw a ship a-sailing
Four and twenty tailors
A little cock sparrow sat on a green tree
Pussy cat, pussy cat
Baa, baa, black sheep
I had a little hen
Diddlety, diddlety, dumpty
Barber, barber, shave a pig
I love little pussy
Hickety, pickety, my black hen
Mary had a little lamb
Ding, dong, bell
Sing, sing, what shall I sing?
Cock-a-doodle-doo !
A carrion crow sat on an oak
Sing a song of sixpence
The north wind doth blow
Hickory, dickory, dock
Three blind mice
A wise old owl lived in an oak
Hey diddle diddle
To market, to market, to buy a fat pig
Great A, little a

The lion and the unicorn
Were fighting for the crown;
The lion beat the unicorn
All round about the town.

Some gave them white bread,
  And some gave them brown;
Some gave them plum-cake,
  And drummed them out of town.

Ladybird, ladybird, fly away home;
    Your house is on fire, your children are gone—
All but one, and her name is Ann,
    And she crept under the frying-pan.

I had a little pony,
    His name was Dapple-grey;
I lent him to a lady,
    To ride a mile away.
She whipped him and she slashed him,
    She rode him through the mire;
I would not lend my pony now
    For all that lady's hire.

Leg over leg,
    As the dog went to Dover,
When he came to a stile,
    Jump he went over.

Little Robin Redbreast sat upon a tree;
Up went pussy cat, down flew he,
Down came pussycat, away Robin ran;
Said little Robin Redbreast, Catch me if you can.
Little Robin Redbreast jumped upon a wall,
Pussy cat jumped after him, and almost had a fall,
Little Robin chirped and sang, What did pussy say?
Pussy said Meow, and Robin flew away.

Goosey, goosey, gander,
  Whither shall I wander?
Upstairs and downstairs,
  And in my lady's chamber.

There I met an old man
    That wouldn't say his prayers;
I took him by the left leg,
    And threw him down the stairs.

I saw a ship a-sailing,
   A-sailing on the sea;
And, oh! it was all laden
   With pretty things for thee.

There were comfits in the cabin,
   And apples in the hold,
The sails were made of silk,
   And the masts of beaten gold.

The four and twenty sailors,
   That stood between the decks,
Were four and twenty white mice
   With chains about their necks.

The captain was a duck,
   With a packet on his back,
And when the ship began to move
   The captain said Quack! Quack!

Four and twenty tailors went to kill a snail;
　The best man among them durst not touch her tail.
She put out her horns like a little Kyloe cow;
　Run, tailors, run, or she'll kill you all e'en now.

A little cock sparrow sat on a green tree,
　And he chirruped, he chirruped, so merry was he.
A naughty boy came with his wee bow and arrow,
　Determined to shoot this little cock sparrow.
This little cock sparrow shall make me a stew,
　And his giblets shall make me a little pie too;
Oh, no, said the sparrow, I *won't* make a stew;
　So he flapped his wings, and away he flew.

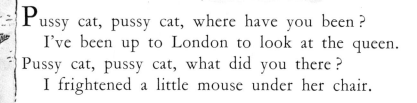

Pussy cat, pussy cat, where have you been?
    I've been up to London to look at the queen.
Pussy cat, pussy cat, what did you there?
    I frightened a little mouse under her chair.

Baa, baa, black sheep,
 Have you any wool?
Yes, sir, yes, sir,
 Three bags full:
One for my master,
 And one for my dame,
And one for the little boy
 Who lives in the lane.

I had a little hen, the prettiest ever seen;
She washed me the dishes, and kept the house clean;
She went to the mill to fetch me some flour,
And brought it home in less than an hour;
She baked me my bread, she brew'd me my ale,
She sat by the fire and told many a fine tale.

Diddlety, diddlety, dumpty,
  The cat ran up the plum tree;
Half-a-crown
To fetch her down,
Diddlety, diddlety, dumpty.

Barber, barber, shave a pig,
  How many hairs will make a wig?
Four-and-twenty, that's enough:
  Give the barber a pinch of snuff.

I love little pussy, her coat is so warm,
  And if I don't hurt her she'll do me no harm.
So I'll not pull her tail nor drive her away,
  But pussy and I very gently will play.

Hickety, pickety, my black hen
She lays eggs for gentlemen.
Gentlemen come every day
To see what my black hen doth lay.

Mary had a little lamb
  Its fleece was white as snow;
And everywhere that Mary went
  The lamb was sure to go.
It followed her to school one day,
  Which was against the rule:
It made the children laugh and play
  To see a lamb at school.
And so the teacher turned it out,
  But still it lingered near,
And waited patiently about
  Till Mary did appear.
What makes the lamb love Mary so?
  The eager children cry,
Why, Mary loves the lamb, you know,
  The teacher did reply.

Ding, dong, bell,
  Pussy's in the well.
Who put her in?
  Little Tommy Lynn.
Who pulled her out?
  Little Johnny Stout.
What a naughty boy was that
  To try to drown poor pussy cat,
Who never did any harm,
  But caught the mice in his father's barn.

Sing, sing, what shall I sing?
  The cat's run off with the pudding-bag string.
Do, do, what shall I do?
  The cat has bitten it quite in two.

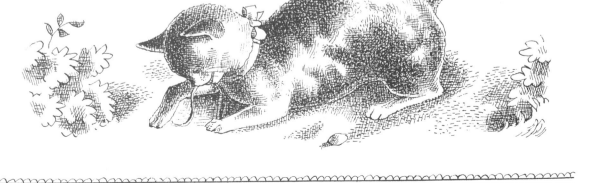

Cock-a-doodle-doo!
  My dame has lost her shoe;
My master's lost his fiddling stick
  And don't know what to do.

Cock-a-doodle-doo!
  What is my dame to do?
Till master finds his fiddling-stick
  She'll dance without her shoe.

Cock-a-doodle-doo !
    My dame has found her shoe,
And master's found his fiddling-stick ;
    Sing doodle-doodle-doo !

Cock-a-doodle-doo !
    My dame will dance with you,
While master fiddles his fiddling stick,
    For dame and doodle-doo.

A carrion crow sat on an oak,
  Watching a tailor shape a cloak.
Sing heigh ho, the carrion crow,
  Fol de riddle, lol de riddle, hi ding do.

Wife bring me my old bent bow,
  That I may shoot yon carrion crow,
Sing heigh ho, the carrion crow,
  Fol de riddle, lol de riddle, hi ding do.

The tailor shot, but he missed his mark,
    And shot his old sow right through the heart.
Sing heigh ho, the carrion crow,
    Fol de riddle, lol de riddle, hi ding do.

Wife bring me brandy in a spoon,
    For our old sow is in a swoon.
Sing heigh ho, the carrion crow,
    Fol de riddle, lol de riddle, hi ding do.

Sing a song of sixpence,
  A pocket full of rye,
Four and twenty blackbirds
  Baked in a pie.
When the pie was open'd,
  The birds began to sing,
Was not that a dainty dish
  To set before the king?

The king was in his counting-house
Counting out his money;
The queen was in the parlour,
Eating bread and honey;

The maid was in the garden,
Hanging out the clothes,
When down flew a blackbird,
And pecked off her nose.
But there came a Jenny Wren
And popped it on again.

The north wind doth blow,
  And we shall have snow,
And what will poor robin do then, poor thing?

He'll sit in a barn,
  And keep himself warm,
And hide his head under his wing, poor thing.

Hickory, dickory, dock,
  The mouse ran up the clock.
The clock struck one,
  The mouse ran down,
Hickory, dickory, dock.

Three blind mice,
See how they run!

They all ran after the farmer's wife,
Who cut off their tails with a carving-knife;

Did you ever hear such a thing in your life?
Three blind mice.

A wise old owl lived in an oak;
The more he saw the less he spoke.
The less he spoke the more he heard:
Why can't we all be like that wise old bird?

Hey diddle diddle,
  The cat and the fiddle,
The cow jumped over the moon;
  The little dog laugh'd
  To see such sport,
And the dish ran away with the
                spoon.

To market, to market, to buy a fat pig;
  Home again, home again, dancing a jig.
To market, to market, to buy a fat hog;
  Home again, home again, jiggety-jog.

Great A, little a,
Bouncing B !
The cat's in the cupboard,
And she can't see me.

A Apple Pie

and

A was an Archer

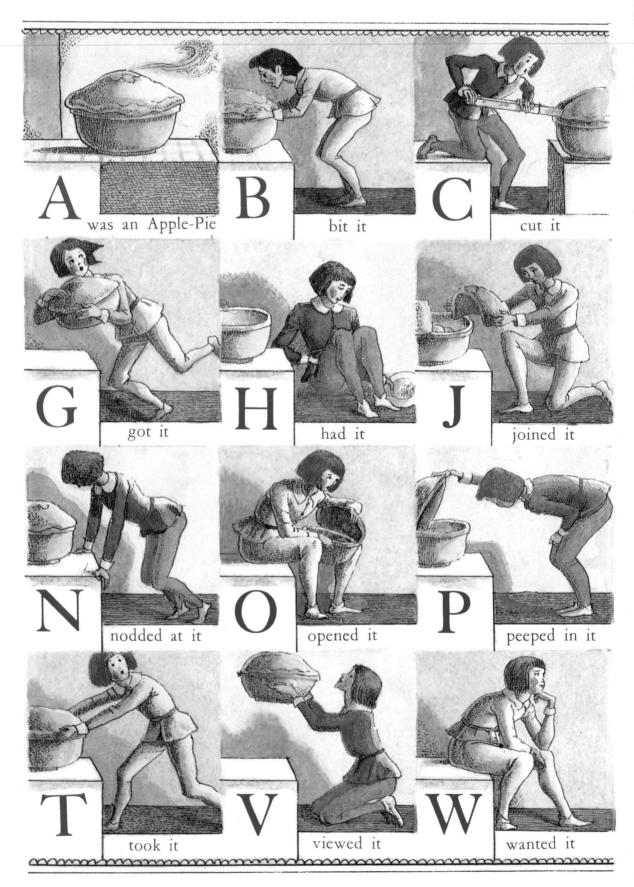

A was an Apple-Pie

B bit it

C cut it

G got it

H had it

J joined it

N nodded at it

O opened it

P peeped in it

T took it

V viewed it

W wanted it

D dealt it

E eat it

F fought for it

K kept it

L longed for it

M mourned for it

Q quartered it

R ran for it

S stole it

X
Y
Z
& ampersand   All wished for a piece in hand.

A was an Archer,
and shot at a frog,

B was a Butcher,
and had a great dog,

C was a Captain,
all covered with lace,

D was a Drummer,
and had a red face,

E was an Esquire,
with pride on his brow,

F was a Farmer,
and followed the plough,

G was a Gamester
who had but ill luck,

H was a Hunter,
and hunted a buck,

I was an Innkeeper
who loved to carouse,

J was a Joiner,
and built up a house,

K was a King,
so mighty and grand,

L was a Lady
who had a white hand,

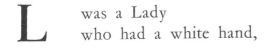

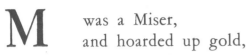

M was a Miser,
and hoarded up gold,

N was a Nobleman,
gallant and bold,

O was an Oyster wench,
and went about town,

P was a Parson,
and wore a black gown,

Q was a Queen
who was fond of good flip,

R was a Robber,
and wanted a whip,

S was a Sailor
who spent all he got,

T was a Tinker
who mended a pot,

U was an Usurer,
a miserable elf,

V was a Vintner
who drank all himself,

W was a Watchman,
and guarded the door,

X was Expensive,
and so became poor,

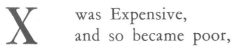

Y was a Youth
that did not love school,

Z was a Zany,
a poor harmless fool.

One, Two, Buckle my Shoe

One, two, buckle my shoe
One to make ready
Mena deena
One I love
On the first day of Christmas
I love sixpence
One, two, three, four, Mary at the cottage door
One, two, three, four, five, once I caught a fish alive

1
2
Buckle my shoe;

3
4
Knock at the door;

5
6
Pick up sticks;

7
8
Lay them straight;

9
10
A good fat hen.

11 12
Dig and delve;

13 14
Maids a-courting;

15 16
Maids in the kitchen;

17 18
Maids a-waiting;

19 20
My plate's empty.

1

One to make ready,
  And two to prepare:
Good luck to the rider,
  And away goes the mare.

2

Mena, deena, deina, duss,
  Catala, weena, weina, wuss,
Spit, spot, must be done,
  Twiddlum, twaddlum, twenty-one!
OUT spells out and out you must go.

One I love, two I love,
Three I love, I say,
Four I love with all my heart,
Five I cast away;
Six he loves, seven she loves, eight both love,
Nine he comes, ten he tarries,
Eleven he courts, twelve he marries.

On the first day of Christmas,
  my true love sent to me ——————

On the second day of Christmas, *etc.* ——
On the third day of Christmas, *etc.* ——
On the fourth day of Christmas, *etc.* ——

   Four colly birds, ◄——
   Three French hens, ◄——
  Two turtle doves and ◄——
 A partridge in a pear-tree. ◄——

On the fifth day of Christmas,
  my true love sent to me

On the sixth day of Christmas, *etc.*
On the seventh day of Christmas, *etc.*
On the eighth day of Christmas, *etc.*

Eight maids a-milking,
Seven swans a-swimming,
Six geese a-laying,
Five gold rings, four colly birds,
Three French hens, two turtle doves and
A partridge in a pear-tree.

On the ninth day of Christmas,
  my true love sent to me

On the tenth day of Christmas, *etc.*
On the eleventh day of Christmas, *etc.*
On the twelfth day of Christmas, *etc.*

Twelve lords a-leaping,
Eleven ladies dancing,
Ten pipers piping,
Nine drummers drumming,
Eight maids a-milking,
Seven swans a-swimming, six geese a-laying,
Five gold rings, four colly birds,
Three French hens, two turtle doves and
A partridge in a pear-tree.

I love sixpence, pretty little sixpence,
　　I love sixpence better than my life;
I spent a penny of it, I lent a penny of it,
　　And I took fourpence home to my wife.

Oh, my little fourpence, pretty little fourpence,
　　I love fourpence better than my life;
I spent a penny of it, I lent a penny of it,
　　And I took twopence home to my wife.

Oh, my little twopence, my pretty little twopence,
    I love twopence better than my life;
I spent a penny of it, I lent a penny of it,
    And I took nothing home to my wife.

Oh, my little nothing, my pretty little nothing,
    What will nothing buy for my wife?
I have nothing, I spend nothing,
    I love nothing better than my wife.

Mary at the cottage door;

Eating cherries off a plate.

1 2 3 4 5 6 7 8 9 10

Once I caught a fish alive.

But I let it go again.

Why did you let it go?
  Because it bit my finger so.
Which finger did it bite?
  The little finger on the right.

Here we go Round the Mulberry Bush

Here we go round the mulberry bush
The High Skip
See saw, Margery Daw
Pease-porridge hot
Cobbler, cobbler, mend my shoe
Oranges and lemons
I sent a letter to my love
Draw a pail of water
Here we come gathering nuts in May

Here we go round the mulberry bush,
  The mulberry bush, the mulberry bush,
Here we go round the mulberry bush,
  All on a frosty morning.

This is the way we clap our hands,
  Clap our hands, clap our hands,
This is the way we clap our hands,
  All on a frosty morning.

This is the way we wash our clothes,
  Wash our clothes, wash our clothes,
This is the way we wash our clothes,
  All on a frosty morning.

The High Skip,
The Sly Skip,
The Skip like a Feather,
The Long Skip,
The Strong Skip,
And the Skip All Together!

The Slow Skip,
The Toe Skip,
The Skip Double-Double,
The Fast Skip,
The Last Skip,
And the Skip Against Trouble!

See, saw, Margery Daw,
　　Johnny shall have a new master;
He shall have but a penny a day,
　　Because he can't work any faster.

Pease-porridge hot,
　　Pease-porridge cold,
Pease-porridge in the pot,
　　Nine days old.
Some like it hot,
　　Some like it cold,
Some like it in the pot,
　　Nine days old.

Cobbler, cobbler, mend my shoe,
Get it done by half-past two,
Half-past two is far too late,
Get it done by half-past eight.

Gay go up and gay go down,
To ring the bells of London Town.
Bull's eyes and targets,
Say the bells of St Margaret's.
Brickbats and tiles,
Say the bells of St Giles'.
Halfpence and farthings,
Say the Bells of St Martin's.
Oranges and lemons,
Say the bells of St Clement's.
Pancakes and fritters,
Say the bells of St Peter's.
Two sticks and an apple,
Say the bells at Whitechapel.
Old Father Baldpate,
Say the slow bells at Aldgate.

Pokers and tongs,
Say the bells of St John's.
Kettles and pans,
Say the bells of St Anne's.
You owe me ten shillings,
Say the bells of St Helen's.
When will you pay me?
Say the bells at Old Bailey.
When I grow rich,
Say the bells at Shoreditch.
Pray when will that be?
Say the bells of Stepney.
I am sure I don't know,
Says the great bell at Bow.
Here comes a candle to light you to bed,
And here comes a chopper to chop off
your head.

I sent a letter to my love
And on the way I dropped it,
A little puppy picked it up
And put it in his pocket.
It isn't you, it isn't you,
But it is you.

Draw a pail of water,
For my lady's daughter.
My father's a king, my mother's a queen,
My two little sisters are dressed in green,
Stamping grass and parsley,
Marigold leaves and daisies.
One rush, two rush,
Prithee, fine lady, come under my bush.

Here we come gathering nuts in May,
Nuts in May, nuts in May,
Here we come gathering nuts in May
On a cold and frosty morning.

Who will you have for nuts in May,
Nuts in May, nuts in May,
Who will you have for nuts in May
On a cold and frosty morning?

The House that Jack Built

The house that Jack built
A farmer went trotting upon his grey mare
Where are you going to, my pretty maid?
Three little kittens
The Old Woman and her pig
Old Mother Hubbard
There was an old woman

This is the house that Jack built.

This is the malt
That lay in the house that Jack built.

This is the rat,
That ate the malt
That lay in the house that Jack built.

This is the cat,
That kill'd the rat,
That ate the malt
That lay in the house that Jack built.

This is the dog,
That worried the cat,
That kill'd the rat,
That ate the malt
That lay in the house that Jack built.

This is the cow
With the crumpled horn,
That toss'd the dog,
That worried the cat,
That kill'd the rat,
That ate the malt
That lay in the house that Jack built.

This is the maiden all forlorn,
That milk'd the cow with the crumpled horn,
That toss'd the dog,
That worried the cat,
That kill'd the rat,
That ate the malt
That lay in the house that Jack built.

This is the man all tatter'd and torn,
That kiss'd the maiden all forlorn,
That milk'd the cow with the crumpled horn,
That toss'd the dog,
That worried the cat,
That kill'd the rat,
That ate the malt
That lay in the house that Jack built.

This is the priest all shaven and shorn,
That married the man all tatter'd and torn,
That kiss'd the maiden all forlorn,
That milk'd the cow with the crumpled horn,
That toss'd the dog,
That worried the cat,
That kill'd the rat,
That ate the malt
That lay in the house that Jack built.

This is the cock that crow'd in the morn,
That wak'd the priest all shaven and shorn,
That married the man all tatter'd and torn,
That kiss'd the maiden all forlorn,
That milk'd the cow with the crumpled horn,
That toss'd the dog,
That worried the cat,
That kill'd the rat,
That ate the malt
That lay in the house that Jack built.

This is the farmer sowing his corn,
That kept the cock that crow'd in the morn,
That wak'd the priest all shaven and shorn,
That married the man all tatter'd and torn,
That kiss'd the maiden all forlorn,
That milk'd the cow with the crumpled horn,
That toss'd the dog,
That worried the cat,
That kill'd the rat,
That ate the malt
That lay in the house that Jack built.

A farmer went trotting
   Upon his grey mare,
Bumpety, bumpety, bump!
With his daughter behind him,
   So rosy and fair,
Lumpety, lumpety, lump.

A raven cried, Croak.
   And they all tumbled down,
Bumpety, bumpety, bump.
The mare broke her knees,
   And the farmer his crown,
Lumpety, lumpety, lump.

The mischievous raven
   Flew laughing away,
Bumpety, bumpety, bump!
And vowed he would serve them
   The same the next day,
Bumpety, bumpety, bump!

Where are you going to, my pretty maid?
I'm going a-milking, sir, she said.
May I go with you, my pretty maid?
You're kindly welcome, sir, she said.
What is your father, my pretty maid?
My father's a farmer, sir, she said.
What is your fortune, my pretty maid?
My face is my fortune, sir, she said.
Then I can't marry you, my pretty maid.
Nobody asked you, sir, she said.

Three little kittens they lost their mittens,
  And they began to cry,
  Oh! mother dear,
  We greatly fear
Our mittens we have lost.

What! lost your mittens, you bad little kittens,
  Then you shall have no pie.
      Mee-ow, mee-ow, mee-ow.
  Then you shall have no pie.

Three little kittens they found their mittens,
  And they began to cry,
  Oh! mother dear,
  See here, see here,
Our mittens we have found.

What! found your mittens, you good little kittens,
  Then you shall have some pie.
      Purr, purr, purr.
  Yes, you shall have some pie.

The three little kittens put on their mittens,
    And soon ate up the pie;
    Oh! mother dear,
    We greatly fear,
Our mittens we have soiled.

What! soiled your mittens, you bad little kittens,
    Then they began to sigh.
        Mee-ow, mee-ow, mee-ow.
    Then they began to sigh.

The three little kittens they washed their mittens,
    And hung them out to dry;
    Oh! mother dear,
    See here, see here,
Our mittens we have washed.

What! washed your mittens, you good little kittens,
    But I smell a rat close by.
        Hush, hush, miew, miew,
    We smell a rat close by,
        Miew, miew, miew.

An old woman was sweeping her house, and she found a little crooked sixpence. What, said she, shall I do with this crooked sixpence? I will go to market and buy a little pig.

As she was coming home, she came to a stile; but the pig would not go over the stile.

She went a little farther, and she met a dog. So she said to the dog —

    Dog, dog, bite pig!
    Pig won't get over the stile,
    And I shan't get home to-night.

But the dog would not.

She went a little farther, and she met a stick. So she said —

    Stick, stick, beat dog!
    Dog won't bite pig;
    Pig won't get over the stile,
    And I shan't get home to-night.

But the stick would not.

She went a little farther, and she met a fire.
So she said —

> Fire, fire, burn stick !
> Stick won't beat dog ;
> Dog won't bite pig ;
> Pig won't get over the stile,
> And I shan't get home to-night.

But the fire would not.
She went a little farther, and she met some water.
So she said —

> Water, water, quench fire !
> Fire won't burn stick ;
> Stick won't beat dog ;
> Dog won't bite pig ;
> Pig won't get over the stile,
> And I shan't get home to-night.

But the water would not.
She went a little farther, and she met an ox.
So she said —

> Ox, ox, drink water !
> Water won't quench fire ;
> Fire won't burn stick ;
> Stick won't beat dog ;
> Dog won't bite pig ;
> Pig won't get over the stile,
> And I shan't get home to-night.

But the ox would not.

She went a little farther, and she met a butcher.
So she said —

> Butcher, butcher, kill ox!
> Ox won't drink water;
> Water won't quench fire;
> Fire won't burn stick;
> Stick won't beat dog;
> Dog won't bite pig;
> Pig won't get over the stile,
> And I shan't get home to-night.

But the butcher would not.

She went a little farther, and she met a rope.
So she said —

> Rope, rope hang butcher!
> Butcher won't kill ox;
> Ox won't drink water;
> Water won't quench fire;
> Fire won't burn stick;
> Stick won't beat dog;
> Dog won't bite pig;
> Pig won't get over the stile,
> And I shan't get home to-night.

But the rope would not.

She went a little farther, and she met a rat.
So she said —

> Rat, rat, gnaw rope!
> Rope won't hang butcher;
> Butcher won't kill ox;
> Ox won't drink water;
> Water won't quench fire;
> Fire won't burn stick;
> Stick won't beat dog;
> Dog won't bite pig;
> Pig won't get over the stile,
> And I shan't get home to-night.

But the rat would not.

She went a little farther, and she met a cat.
So she said —

> Cat, cat, kill rat!
> Rat won't gnaw rope;
> Rope won't hang butcher;
> Butcher won't kill ox;
> Ox won't drink water;
> Water won't quench fire;
> Fire won't burn stick;
> Stick won't beat dog;
> Dog won't bite pig;
> Pig won't get over stile,
> And I shan't get home to-night.

The cat said, If you will give me a saucer of milk,
I will kill the rat.

So the old woman gave the cat the milk, and when
she had lapped up the milk —

> The cat began to kill the rat,
> The rat began to gnaw the rope,
> The rope began to hang the butcher,
> The butcher began to kill the ox,
> The ox began to drink the water,
> The water began to quench the fire,
> The fire began to burn the stick,
> The stick began to beat the dog,
> The dog began to bite the pig,
> The pig jumped over the stile,
> And so the old woman got home that night.

Old Mother Hubbard
Went to the cupboard
    To get her poor dog a bone,
But when she came there
The cupboard was bare,
    And so the poor dog had none.

She went to the baker's
    To buy him some bread,
But when she came back
    The poor dog was dead.

She went to the joiner's
    To buy him a coffin,
But when she came back
    The poor dog was laughing.

She took a clean dish
    To get him some tripe,
But when she came back
    He was smoking his pipe.

She went to the fishmonger's
    To buy him some fish,
But when she came back
    He was licking the dish.

She went to the ale-house
    To get him some beer,
But when she came back
    The dog sat in a chair.

She went to the tavern
    For white wine and red,
But when she came back
    The dog stood on his head.

She went to the hatter's
    To buy him a hat,
But when she came back
    He was feeding the cat.

She went to the barber's
    To buy him a wig,
But when she came back
    He was dancing a jig.

She went to the fruiter's
    To buy him some fruit,
But when she came back
    He was playing the flute.

She went to the tailor's
  To buy him a coat,
But when she came back
  He was riding a goat.

She went to the cobbler's
  To buy him some shoes,
But when she came back
  He was reading the news.

She went to the sempstress
  To buy him some linen,
But when she came back
  The dog was a-spinning.

She went to the hosier's
  To buy him some hose,
But when she came. back
  He was dressed in his clothes.

The dame made a curtsey,
  The dog made a bow;
The dame said, Your servant.
  The dog said, Bow wow.

There was an old woman, as I've heard tell,
    She went to market her eggs for to sell;
She went to market all on a market-day,
    And she fell asleep on the king's highway.

There came by a pedlar whose name was Stout;
    He cut her petticoats all round about;
He cut her petticoats up to the knees,
    Which made the old woman to shiver and sneeze.

When this little woman first did wake,
 She began to shiver and she began to shake;
She began to wonder and she began to cry,
 Lackamercyme, this is none of I!

But if it be I, as I do hope it be,
 I've a little dog at home, and he'll know me;
If it be I, he'll wag his little tail,
 And if it be not I, he'll loudly bark and wail.

Home went the little woman all in the dark;
 Up got the little dog and he began to bark;
He began to bark, so she began to cry,
 Lackamercyme, this is none of I!

If Ifs and Ands

If ifs and ands
If wishes were horses
When the wind is in the east
A swarm of bees in May
March winds
A sunshiny shower
Rain, rain, go to Spain
St Swithin's Day
The fifth of November
See a pin and pick it up
For every evil under the sun
Winter's thunder
A red sky at night
Early to bed
They that wash on Monday
Monday's child
Cuckoo, cuckoo, what do you do?
Sneeze on Monday
Thirty days hath September
For want of a nail
Two legs
As I was going to St. Ives
Four stiff-standers
As soft as silk
Old Mother Twitchett
Little Nancy Etticoat
A house full, a yard full
Long legs, crooked thighs
I have a little sister
Lilies are white

If ifs and ands were pots and pans,
There'd be no work for tinkers' hands.

If wishes were horses
   Beggars would ride:
If turnips were watches
   I'd wear one by my side.

When the wind is in the east,
   'Tis neither good for man nor beast;
When the wind is in the north,
   The skilful fisher goes not forth;
When the wind is in the south,
   It blows the bait in the fishes' mouth;
When the wind is in the west,
   Then 'tis at the very best.

A swarm of bees in May
   Is worth a load of hay;
A swarm of bees in June
   Is worth a silver spoon;
A swarm of bees in July
   Is not worth a fly.

March winds and April showers
   Bring forth May flowers.

A sunshiny shower
  Won't last half an hour.

Rain, rain, go to Spain,
And never, never, never
Come back again.

St Swithin's Day, if thou dost rain
  Forty days it will remain:
St Swithin's Day, if thou be fair
  For forty days 'twill rain nae mair.

Please to remember the fifth of November
Gunpowder, treason and plot
I see no reason why gunpowder treason
Should ever be forgot.

See a pin and pick it up,
  All the day you'll have good luck:
See a pin and let it lay,
  Bad luck you'll have all day!

For every evil under the sun
There is a remedy, or there is none.
If there be one, try and find it;
If there be none, never mind it.

Winter's thunder
Is the world's wonder.

A red sky at night is the shepherd's delight,
  A red sky in the morning is the shepherd's warning.

Early to bed and early to rise,
Makes a man healthy, wealthy, and wise.

The cock doth crow
To let thee know,
An thou be wise,
'Tis time to rise.

They that wash on Monday
    Have all the week to dry,
They that wash on Tuesday
    Are not so much awry,
They that wash on Wednesday
    Are not so much to blame,
They that wash on Thursday,
    Wash for shame,
They that wash on Friday,
    Wash in need,
But they that wash on Saturday,
    Oh! they're sluts indeed.

Monday's child is fair of face,
  Tuesday's child is full of grace,
Wednesday's child is full of woe,
  Thursday's child has far to go,
Friday's child is loving and giving,
  Saturday's child works hard for its living,
And the child that is born on the Sabbath day
  Is bonny and blithe, and good and gay.

Cuckoo, cuckoo, what do you do?
In April come I will,
In May I sing all day,
In June I change my tune,
In July I prepare to fly,
In August go I must.

Sneeze on Monday, sneeze for danger;
Sneeze on Tuesday, kiss a stranger;
Sneeze on Wednesday, get a letter;
Sneeze on Thursday, something better;
Sneeze on Friday, sneeze for sorrow;
Sneeze on Saturday, see your true love
to-morrow.

Thirty days hath September,
April, June and November;
All the rest have thirty-one,
Except February alone,
Which has twenty-eight days clear
And twenty-nine in each leap-year.

For want of a nail, the shoe was lost,
For want of a shoe, the horse was lost,
For want of a horse, the rider was lost,
For want of a rider, the battle was lost,
For want of a battle, the kingdom was lost,
And all for the want of a horseshoe nail.

Two legs sat upon three legs
   With one leg in his lap;
In comes four legs,
   Runs away with one leg,
Up jumps two legs,
   Picks up three legs,
Throws it after four legs,
   And makes him bring back one leg.

As I was going to St Ives,
   I met a man with seven wives,
Each wife had seven sacks,
   Each sack had seven cats,
Each cat had seven kits;
   Kits, cats, sacks, and wives,
How many were going to St Ives?

Four stiff-standers,
Four dilly-danders,
Two lookers, two crookers
And a wig-wag.

As soft as silk,
As white as milk,
As bitter as gall,
A thick wall
And a green coat covers me all.

TO
ST
IVES

Old Mother Twitchett had but one eye,
And a long tail which she let fly;
Every time she went over a gap,
She left a bit of her tail in a trap.

Little Nancy Etticoat
   With a white petticoat,
And a red nose;
   The longer she stands
The shorter she grows.

A house full, a yard full,
   And ye can't catch a bowl full.

Long legs, crooked thighs,
   Little head, and no eyes.

I have a little sister,
   They call her Pretty Peep;
She wades in the waters
   Deep, deep, deep,
She climbs the mountains
   High, high, high;
My poor little sister,
   She has but one eye.

Lilies are white,   Roses are red,
Rosemary's green.   Lavender's blue.
When you are King,   If you will have me,
I will be Queen.   I will have you.

Cradle Rhyme Games

Pat-a-cake, pat-a-cake, baker's man

Often the first little action game taught to a baby. Clap hands together, pat an imaginary lump of dough and throw it away into the oven. The precise actions vary in nearly every family.

This little pig went to market

Played with fingers or toes — but much more fun with bare toes! Beginning with the big toe, each in turn is a little pig.

Walk down the path

Walk fingers across head, knock on forehead, tweak nose, brush fingers on chin and pop finger into open mouth.

Ring-a-ring o' roses

Baby's first round game. Take hands, facing inwards, and walk round in a circle and fall down on the last line.

Shoe the little horse

Tap baby's bare foot.

Round and round the cornfield

Hold the child's hand palm upwards with one hand, and with fingers of the other trace circle round palm, creeping up arm to neck on last line.

Can you keep a secret?

Tickle child's hand or foot — the child tries not to laugh.

This is the way the ladies ride

Give child a 'ride' on lap or knees.

Ride a cock-horse

Bounce child on knee or, holding its hands, put the child facing you astride one foot.

Dance, Thumbkin, dance

Hold up one hand. Make thumb move alone, then all fingers; and so on with Foreman, Longman and Ringman, but Littleman *cannot* dance alone.

Here sits the Lord Mayor

Lord Mayor	touch the forehead
his men	„ „ two eyes
the cock	„ one cheek
the hen	„ the other cheek
little chicken	„ the nose
here they run	run in the open mouth
chinchopper	chuck under chin

Here are my lady's knives and forks

1. Hands back to back — fingers outstretched and interlocked.
2. With turn of wrist bring back of hands, fingers still interlocked, on top. Flatten knuckles to make table top — thumbs tucked in.
3. Bring little fingers up, tips touching, to make mirror at back of table.
4. Bring up fore-fingers, tips touching, to make high head and foot of cradle.

Here's a ball for baby

1. Palms facing, join thumb and finger tips to make a round.
2. Clench both fists and 'hammer' one on the other.
3. Hold both hands up, fingers straight.
4. Clap hands together.
5. One hand out flat, palm downwards — for umbrella top. Extended forefinger of other hand as handle.
6. Palms upwards, fingers interlocked across backs to make hollow cradle.

Two little dicky birds

Stick paper to, or otherwise mark, the nails or back of each forefinger in preparation for this elementary 'sleight-of-hand'.
Put forefingers on table or any firm surface:

one is Peter — one is Paul.

On 'fly away' lift hands shoulder-high and, in mid-air, quickly extend middle fingers in place of forefingers before bringing hands down again. Repeat, extending forefingers, for 'come back'.

Nursery Rhyme Games

Pease-porridge hot

There seem to be several ways of clapping in time to this. Here is one version:

Two children stand close together facing.

Pease	— clap both hands on own thighs (arms straight)
Porridge	— clap own hands together
Hot	— clap hands with opposite (a quick turn of wrist to bring palms facing outwards)
Pease-porridge cold	— repeat above
Pease	— clap own hands together
Porridge	— right hand diagonally with opposite's right hand
In	— clap own hands
The	— left hand diagonally with opposite's left hand
Pot	— clap own hands
Nine days old	— repeat first part of first movement

Here we go round the mulberry bush

The first verse is sung while children holding hands and facing inwards dance round in a circle. Succeeding verses are sung, with hands free for acting, by the children standing in place or following each other in a circle, walking, skipping, etc. The first verse is often repeated between each action verse.

Here we come gathering nuts in May

Children first pick up sides, and standing in two lines facing each other they hold hands and advance and retreat in turn singing:

1st line — Here we come gathering nuts in May, etc.
2nd line — Who will you have for nuts in May, etc.
1st line — (having decided amongst themselves) We'll have *child's name* for nuts in May, etc.
2nd line — Who will you send to fetch her (him) away, etc.
1st line — We'll send .... to fetch her (him) etc.

The two named children advance and, putting the toes of their right feet at a crack or otherwise indicated line, each tries to pull the other across with the right hand. The successful one takes his opponent across to his side and the game continues until one side has won all members of the opposing side.

I sent a letter to my love

All the children except one join hands in as big a circle as they can make, facing inwards. The 'odd man out' walks round outside the circle, singing the rhyme. On 'it is you' he drops the handkerchief behind a child, who must then pick it up and race round in the opposite direction trying to reach his place again before the first child gets there. Whoever is left out continues the game as before.

Oranges and lemons

Two children (usually the tallest), secretly deciding beforehand which is to be Oranges and which Lemons, make an arch with their joined hands. The rest of the children walk under the arch and round. On the last line a child is caught by lowering of the arch and is asked privately to choose oranges or lemons. He then stands behind whichever he chooses. And so the game continues till all players are captured. The two sides then have a tug-of-war.

Draw a pail of water

This game is similar to Oranges and Lemons. The free children sing the verse, and the two making the arch sing the last two lines.

Hunt the slipper

All children except one sit cross-legged on the floor in a ring facing inwards, one having concealed a slipper under himself. The odd man out comes into the ring and tries to find out who has the slipper by asking first one and then another. The slipper can be passed from child to child.

Index of First Lines

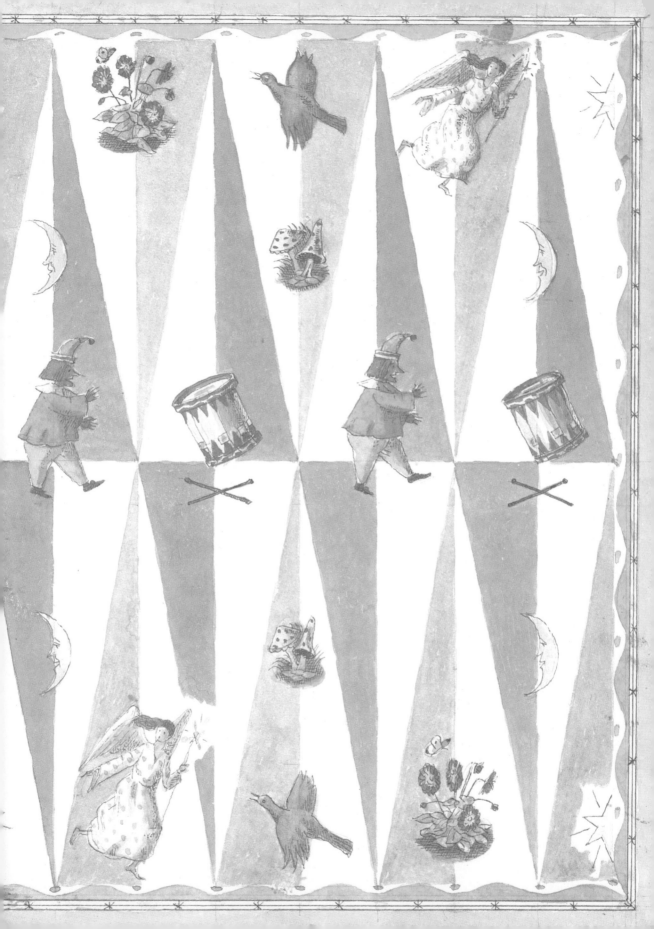